Three studies in Computational Complexity Theory

Valentin B. Bura

The Polynomial Hierarchy Collapses

Valentin Bura[a]

[a]*Department of Computer Science, University of Liverpool L69 3BX*

Abstract

This paper illustrates the power of Gaussian Elimination by adapting it to the problem of Exact Satisfiability. For 1-in-3 SAT instances with non-negated literals we are able to obtain considerably smaller equivalent instances of 0/1 Integer Programming restricted to equality only.

Thus we obtain an upper bound for the complexity of its counting version of $\mathcal{O}(2\kappa r 2^{(1-\kappa)r})$ for number of variables r and clauses-to-variables ratio κ. Combining this method with previous results gives a time and space complexity for the counting problem of $\mathcal{O}(4/3|V|2^{3|V|/8})$ and $\mathcal{O}(4/3|V|2^{3|V|/16})$.

Our method shows that Positive instances of 1-in-3 SAT may be reduced to significantly smaller instances of I.P. in the following sense: any such instance of $|V|$ variables and $|C|$ clauses can be polynomial-time reduced to an instance of 0/1 Integer Programming with equality only, of size at most $2/3|V|$ variables and at most $|C|$ clauses.

We then proceed to define formally the notion of a non-trivial kernel. For this, we define the problems considered as Constraint Satisfaction Problems. Considering recent advances in Computational Complexity relating to sparsification and existence of non-trivial kernels, we conclude by showing that the method presented here, giving a non-trivial kernel for positive 1-in-3 SAT, implies the existence of a non-trivial kernel for 1-in-3 SAT.

Our proof shows the structure known as the Polynomial Hierarchy collapses to the level above $P = NP$.

Keywords: Computational Complexity, Boolean Satisfiability, Kernelization

1. Introduction

Recall that `SAT` and its restrictions `cnf-SAT`, `k-cnf-SAT` and `3-cnf-SAT` are `NP-complete` as shown in [1, 2, 3]. The `1-3-SAT` problem is that, given a collection of triples over some variables, to determine whether there exists a truth assignment to the variables so that each triple contains exactly one true literal and exactly two false literals.

Schaefer's reduction given in [4] transforms an instance of `3-cnf-SAT` into a `1-3-SAT` instance. A simple truth-table argument shows this reduction to be parsimonious, hence `1-3-SAT` is complete for the class #P while a parsimonious reduction from `1-3-SAT` also shows `1-3-SAT`$^+$ to be complete for #P.

The `1-K-SAT` problem, a generalization of `1-3-SAT`, is that, given a collection of tuples of size K over some variables, to determine whether there exists a truth assignment to the variables so that each K-tuple contains exactly one true and $K-1$ false literals.

The `1-K-SAT` problem has been studied before under the name of XSAT. In [5] very strong upper bounds are given for this problem, including the counting version. These bounds are $\mathcal{O}(1.1907^{|V|})$ and $\mathcal{O}(1.2190^{|V|})$ respectively, while in [6] the same bound of $2^{|C|}|V|^{\mathcal{O}(1)}$ is given for both decision and counting, where $|V|$ is the number of variables and $|C|$ the number of clauses.

Gaussian Elimination was used before in the context of boolean satisfiability. In [7] the author uses this method for handling xor types of constraints. Other recent examples of Gaussian elimination used in exact algorithms or kernelization may be indeed found in the literature [8, 9].

Hence the idea that constraints of the type implying this type of exclusivity can be formulated in terms of equations, and therefore processed using Gaussian Elimination, is not new and the intuition behind it is very straightforward.

We mention the influential paper by Dell and Van Melkebeek [10] together with a continuation of their study by Jansen and Pieterse [11, 12]. It is shown in these papers that, under the assumption that coNP $\not\subseteq$ NP \ poly, there cannot exist a significantly small kernelization of various problems, of which exact satisfiability is one. We shall use these results directly in our current approach.

We begin our investigation by showing how a 1-3-SAT$^+$ instance can be turned into an integer programming version 0-1-IP$^=$ instance with fewer variables. The number of variables in the 0-1-IP$^=$ instance is at most two-thirds of the number of variables in the 1-3-SAT$^+$ instance. We achieve this by a straightforward preprocessing of the 1-3-SAT$^+$ instance using Gauss-Jordan elimination.

We are then able to count the solutions of the 1-3-SAT$^+$ instance by performing a brute-force search on the 0-1-IP$^=$ instance. This method gives interesting upper bounds on 1-3-SAT$^+$, and the associated counting problem, though without a further analysis, the bounds thus obtained may not be the strongest upper bounds found in the literature for these problems.

Our method shows how instances become easier to solve with variation in clauses-to-variables ratio. For random k-cnf-SAT the ratio of clauses to variables has been studied intensively, for example [13] gives the proof that a formula with density below a certain threshold is with high probability satisfiable while above the threshold is unsatisfiable.

The ratio plays a similar role in our treatment of 1-3-SAT. Another important observation is that in our case this ratio cannot go below 1/3 up to uniqueness of clauses, at the expense of poly-time pre-processing of the problem instance. We note that, by reduction from 3-cnf-SAT any instance of 1-3-SAT in which the number of clauses does not exceed the number of variables is also NP-complete. Hence we restrict our attention to these instances.

Our preprocessing induces a certain type of "order" on the variables, such that some of the non-satisfying assignments can be omitted by our solution search. We therefore manage to dissect the 1-K-SAT instance and obtain a "core" of variables on which the search can be performed. For a treatment of Parameterized Complexity the reader is directed to [14].

2. Outline

After a brief consideration of the notation used in Section 3, we define in Section 4 the problems 1-3-SAT, 1-3-SAT$^+$ and the associated counting problems #1-3-SAT and #1-3-SAT$^+$. We elaborate on the relationship between the number of clauses and the number of variables in 1-3-SAT$^+$. We give a proof sketch that 1-3-SAT is NP − complete via a reduction from 3-cnf-SAT, and a proof sketch that 1-3-SAT$^+$ is NP − complete via reduction from 1-3-SAT. We conclude by remarking that due to this chain of reductions, the restriction of 1-3-SAT$^+$ to instances with more variables than clauses is also NP − complete, since these kind of instances encode the 3-cnf-SAT problem. We hence restrict our treatment of 1-3-SAT$^+$ to these instances.

Section 5 presents our method of reducing a 1-3-SAT$^+$ instance to an instance of 0/1 Integer Programming with Equality only. This results in a 0/1 I.P.E. instance with at most two thirds the number of variables found in the 1-3-SAT$^+$ instance. Essentially, our method describes the same method as the one presented by Jansen and Pieterse in an introductory paragraph of [12]. Jansen and Pieterse are not however primarily interested in reduction of the number of variables, but reduction of number of constraints and they do not tackle the associated counting problem as such. The method consists of encoding a 1-3-SAT$^+$ instance into a system of linear equations and performing Gaussian Elimination on it.

Linear Algebraic methods show the resulting matrix can be rearranged into an $r \times r$ diagonal submatrix of "independent" columns, where r is the rank of the system, to which it is appended a submatrix containing the rest of the columns and the result column which correspond roughly to the 0/1 I.P.E. instance we have in mind. We further know the values in the independent submatrix can be scaled to 1.

The most pessimistic scenario complexity-wise is when the input clauses, or the rank of the resulting system, is a third the number of variables, $|C| = 1/3|V|$, from which we obtain our complexity upper bounds.

To this case, one may wish to contrast the case of the system matrix being full rank, for which Gaussian Elimination alone suffices to find a solution. Further to this, we explain how to solve the 0/1 I.P.E. problem in order to recover the number of solutions to the 1-3-SAT$^+$ problem.

Section 6 outlines the complexity implications for 1-3-SAT$^+$ by considering the 1981 results of Schroeppel and Shamir [15]. Section 7 outlines the method of substitution, well-known to be equivalent to Gaussian Elimination.

Finally, Section 8 gives additional definitions, considers some of the very recent literature on sparsification, and gives an argument that the existence of the 1-3-SAT$^+$ kernel found in previous sections implies the existence of a non-trivial kernel for the more general 1-3-SAT.

3. Notation

We denote boolean variables by $p_1, p_2, \ldots, p_i, \ldots$ and denote negation by $\neg p_i$. Whenever considered as binary variables over the set $\{0, 1\}$ these will be written as $\bar{p}_i$ in the positive case and $-\bar{p}_i$ in the negative.

Denote the true and false constants by $\top$ and $\bot$ respectively. For any SAT formula φ, write $\Sigma(\varphi)$ if φ is satisfiable and write $\bar{\Sigma}(\varphi)$ otherwise. Reserve the notation $a(p)$ for a truth assignment to the variable p.

We write $\Phi(r,k)$ for the set of formulas in 3-CNF with r variables and k unique clauses. We also write $\varphi(V,C)$ to specify concretely such a formula, where V, C shall denote the sets of variables and clauses of φ. For any formula $\varphi \in \Phi(r,k)$ we let $\kappa(\varphi) = \frac{k}{r}$.

We will make use of the following properties of a given map f:

subadditivity: $f(A+B) \leq f(A) + f(B)$

scalability: $f(cA) = cf(A)$ for constant c.

For a given tuple $s = (s_1, s_2, \ldots, s_n)$ we let $s(m)$ denote the element s_m. Finally, for given linear constraints $\sum_{i \leq n} d_i x_i = R$ for some n and $x_i \in \{0,1\}$, denote by $coef(x_i)$ the value d_i.

4. One-in-Three SAT

One-in-three satisfiability arose in late seventies as an elaboration relating to Schaefer's Dichotomy Theorem [4]. It is proved there using certain assumptions boolean satisfiability problems are either in P or they are NP-complete.

The counting versions of satisfiability problems were introduced in [16] and it is known in general that counting is in some sense strictly harder than the corresponding decision problem.

This is due to the fact that, for example, producing the number of satisfying assignments of a formula in 2-CNF is complete for #P, while the corresponding decision problem is known to be in P [16]. We thus restrict our attention to 1-3-SAT and more precisely 1-3-SAT$^+$ formulas.

Definition 4.1 (1-3-SAT). *1-3-SAT is defined as determining whether a formula $\varphi \in \Phi(r, k)$ is satisfiable, where the formula comprises of a collection of triples*

$$C = \{\{p_1^1, p_2^1, p_3^1\}, \{p_1^2, p_2^2, p_3^2\}, \ldots, \{p_1^k, p_2^k, p_3^k\}\}$$

such that $p_1^i, p_2^i, p_3^i \in V = \{p_1, \neg p_1, p_2, \neg p_2, \ldots, p_r, \neg p_r\} \cup \{\bot\}$ and for any clause exactly one of the literals is allowed to be true in an assignment, and no clause may contain repeated literals or a literal and its negation, and such that every variable in V appears in at least one clause.

In the restricted case that $p_1^i, p_2^i, p_3^i \in V^+ = \{p_1, p_2, \ldots, p_r\} \cup \{\bot\}$ for $1 \leq i \leq r$ we denote the problem as 1-3-SAT$^+$.

In the extended case that we are required to produce the number of satisfying assignments, these problems will be denoted as #1-3-SAT and #1-3-SAT$^+$.

Example 4.1. *The 1-3-SAT$^+$ formula $\varphi = \{\{p_1, p_2, p_3\}, \{p_2, p_3, p_4\}\}$ is satisfiable by the assignment $a(p_2) = \top$ and $a(p_j) = \bot$ for $j = 1, 3, 4$. The 1-3-SAT$^+$ formula $\varphi = \{\{p_1, p_2, p_3\}, \{p_2, p_3, p_4\}, \{p_1, p_2, p_4\}, \{p_1, p_3, p_4\}\}$ is not satisfiable.*

Lemma 4.1. *Up to uniqueness of clauses and variable naming the set $\Phi(r, r/3)$ determines one 1-3-SAT$^+$ formula and this formula is trivially satisfiable.*

Proof. Consider the formula $\varphi = \{\{p_{3i}, p_{3i+1}, p_{3i+2}\} \mid 1 \leq i \leq r/3\}$ which has r variables and $r/3$ clauses, hence belongs to the set $\Phi(r, r/3)$ and it is satisfiable, trivially, by any assignment that makes each clause evaluate to true.

Now take any clause $\{a, b, c\} \in \varphi(V, C)$ with $a, b, c \in V$. We claim there is no other clause $\{a', b', c'\} \in \varphi$ such that $\{a, b, c\} \cap \{a', b', c'\} \neq \emptyset$, for otherwise let a be in their intersection and we can see the number of variables used by the $r/3$ clauses reduces by one variable, to be $r - 1$. Now, since the clauses of φ do not overlap in variables, we can see that our uniqueness claim must hold, since the elements of φ are partitions of the set of variables. □

Remark 4.1. *For* 1-3-SAT$^+$, *the sets* $\Phi(r,k)$ *for* $k < r/3$ *are empty.*

Schaefer gives a polynomial time parsimonious reduction from 3-cnf-SAT to 1-3-SAT hence showing that 1-3-SAT and its counting version #1-3-SAT are NP-complete and respectively #P-complete.

Theorem 4.1 (Schaefer, [4]). 1-3-SAT *is* NP-complete.

Proof sketch:. Proof by reduction from 3-cnf-SAT. For any clause $p \vee p' \vee p''$ create three 1-3-SAT clauses $\{\neg p, a, b\}$, $\{p', b, c\}$, $\{\neg p'', c, d\}$. Hence, we obtain an instance with $|V| + 4|C|$ variables and $3|C|$ clauses, for the instance of 3-cnf-SAT of $|V|$ variables and $|C|$ clauses. □

The following statement is given in [17]. For the sake of completeness, we provide a proof by a parsimonious reduction from 1-3-SAT.

Theorem 4.2 ([17]). 1-3-SAT$^+$ *is* NP-complete.

Proof sketch:. Construct an instance of 1-3-SAT$^+$ from an instance of 1-3-SAT. Add every clause in the 1-3-SAT instance with no negation to the 1-3-SAT$^+$ instance. For every clause containing one negation $\{\neg p, p', p''\}$, add to the 1-3-SAT$^+$ two clauses $\{\hat{p}, p', p''\}$ and $\{\hat{p}, p, \bot\}$ where $\hat{p}$ is a fresh variable. For a clause containing two negations $\{\neg p, \neg p', p''\}$ we add two fresh variables $\hat{p}, \hat{p}'$ and three clauses $\{\hat{p}, \hat{p}', p''\}$, $\{\hat{p}, p, \bot\}$ and $\{\hat{p}', p', \bot\}$. For a clause containing three negations $\{\neg p, \neg p', \neg p''\}$ we add three fresh variables $\hat{p}, \hat{p}', \hat{p}''$ and four clauses $\{\hat{p}, \hat{p}', \hat{p}''\}$, $\{\hat{p}, p, \bot\}$, $\{\hat{p}', p', \bot\}$ and $\{\hat{p}'', p'', \bot\}$. We obtain a 1-3-SAT$^+$ formula with at most $4|C|$ more clauses and at most $|V| + 3|C|$ more variables, for initial number of clauses and variables $|C|$ and $|V|$ respectively. Our reduction is parsimonious, for it is verifiable by truth-table the number of satisfying assignments to the 1-3-SAT clause $\{\neg p, \neg p', \neg p''\}$ is the same as the number of satisfying assignments to the 1-3-SAT$^+$ collection of clauses $\{\{\hat{p}, \hat{p}', \hat{p}''\}, \{\hat{p}, p, \bot\}, \{\hat{p}', p', \bot\}, \{\hat{p}'', p'', \bot\}\}$. □

Remark 4.2. *In virtue of Theorem 4.1 and Theorem 4.2 we restrict ourselves to instances of* 1-3-SAT$^+$ $\varphi \in \Phi(r,k)$ *with* $r \geq k$. *For if an instance of* 3-cnf-SAT $\bar{\varphi} \in \Phi(r',k')$ *is reduced to an instance of* 1-3-SAT $\hat{\varphi} \in \Phi(r'',k'')$ *then our reduction entails* $r'' = r' + 4k'$ *and* $k'' = 3k'$.

We analyze the further reduction to the instance of 1-3-SAT$^+$ $\varphi \in \Phi(r,k)$. *Let* C, C', C'', C''' *be the collections of clauses in* $\hat{\varphi}$ *containing, no negation, one negation, two negations and three negations respectively.*

Our reduction implies $r = r'' + |C'| + 2|C''| + 3|C'''|$ *and* $k = |C| + 2|C'| + 3|C''| + 4|C'''| = k'' + |C'| + 2|C''| + 3|C'''|$. *Then,* $r - k = r'' + |C'| + 2|C''| + 3|C'''| - k'' - |C'| - 2|C''| - 3|C'''| = r'' - k'' = r' + 4k' - 3k' = r' + k' > 0$.

5. Gaussian elimination

A rank function is used as a measure of "independence" for members of a certain set. The dual notion of nullity is defined as the complement of the rank with respect to the size of the set.

Definition 5.1. *A rank function R obeys the following*

1. $R(\emptyset) = 0$,

2. $R(A \cup B) \leq R(A) + R(B)$,

3. $R(A) \leq R(A \cup \{a\}) \leq R(A) + 1$.

Definition 5.2 (rank and nullity). *For a* 1-3-SAT$^+$ *formula* $\varphi(V,C)$ *define the system of linear equations* $\mathtt{Sys}(\varphi)$ *as follows:*

for any clause $\{p, p', p''\} \in C$ *add to* $\mathtt{Sys}(\varphi)$ *equation* $\bar{p} + \bar{p}' + \bar{p}'' = 1$;

Define the rank and nullity of φ *as* $\eta(\varphi) = R(\mathtt{Sys}(\varphi))$ *and* $\bar{\eta}(\varphi) = |V| - \eta(\varphi)$. *If formula is clear from context, we also use the shorthand* η *and* $\bar{\eta}$.

Remark 5.1. η *is a rank function with respect to sets of* 1-3-SAT *triples.*

Lemma 5.1. *For any* 1-3-SAT$^+$ *instance φ transformed into a linear system* Sys(φ) *we observe the following:*

$\bar{p} + \bar{p}' + \bar{p}'' = 1$ *has a solution $S \subset \{0,1\}^3$ if and only if exactly one of $\bar{p}, \bar{p}', \bar{p}''$ is equal to 1 and the other two are equal to 0.*

Proposition 5.1. *For any formula $\varphi \in \Phi(r,k)$ we have $\Sigma(\varphi)$ if and only if* Sys(φ) *has at least one solution over $\{0,1\}^r$.*

Corollary 5.1. *A formula $\varphi \in \Phi(r,k)$ has as many satisfiability assignments as the number of solutions of* Sys(φ) *over $\{0,1\}^r$.*

We define the binary integer programming problem with equality here and show briefly that 1-3-SAT$^+$ is reducible to a "smaller" instance of this problem.

Definition 5.3 (0,1-integer programming with equality). *The* 0-1-IP$^=$ *problem is defined as follows. Given a family of finite tuples $s_1, s_2, \ldots, s_k$ with each $s_i \in \mathbb{Q}^S$ for some fixed $S \in \mathbb{N}$, and given a sequence $q_1, q_2, \ldots, q_k \in \mathbb{Q}$, decide whether there exists a tuple $T \in \{0,1\}^S$ such that*

$$\sum_{j=1}^{S} s_i(j) T(j) = q_i \text{ for each } i \in \{1, 2, \ldots, k\}$$

Remark 5.2. 0-1-IP$^= \in \mathcal{O}(k2^S)$ *where k is the number of* 0-1-IP$^=$ *tuples and S is the size of the tuples.*

Proof sketch:. The bound is obtained through applying an exhaustive search. □

Lemma 5.2. *Let $\varphi \in \Phi(r,k)$ be a* 1-3-SAT$^+$ *formula, then $\eta(\varphi) \leq k$ and $\bar{\eta}(\varphi) \geq r - k$.*

Proof. Follows from the observation that η is a rank function. □

Lemma 5.3. *Consider a* 1-3-SAT$^+$ *formula φ and suppose $\eta(\varphi) = k$ and $\bar{\eta}(\varphi) = r - k$. The satisfiability of φ is decidable in $\mathcal{O}(2k2^{r-k})$.*

Proof. The result of performing Gauss-Jordan Elimination on $\mathtt{Sys}(\varphi)$ yields, after a suitable re-arrangement of column vectors, the reduced echelon form

$$\mathtt{GJE}(\mathtt{Sys}(\varphi)) = \begin{bmatrix} 1 & 0 & 0 & \ldots & 0 & x_{11} & x_{12} & \ldots & x_{1d} & R_1 \\ 0 & 1 & 0 & \ldots & 0 & x_{21} & x_{22} & \ldots & x_{2d} & R_2 \\ 0 & 0 & 1 & \ldots & 0 & x_{31} & x_{32} & \ldots & x_{3d} & R_3 \\ \vdots & \vdots & \vdots & & \vdots & \vdots & \vdots & \vdots & \vdots & \vdots \\ 0 & 0 & 0 & \ldots & 1 & x_{k1} & x_{k2} & \ldots & x_{kd} & R_k \end{bmatrix}$$

Now consider the following structure, obtained from the given dependencies above through ignoring the zero entries

$$\begin{bmatrix} 1 & x_{11} & x_{12} & \ldots & x_{1d} & R_1 \\ 1 & x_{21} & x_{22} & \ldots & x_{2d} & R_2 \\ 1 & x_{31} & x_{32} & \ldots & x_{3d} & R_3 \\ \vdots & \vdots & \vdots & & \vdots & \vdots \\ 1 & x_{k1} & x_{k2} & \ldots & x_{kd} & R_k \end{bmatrix}$$

This induces an instance of 0-1-IP$^=$ which can be solved as follows

---
Initialize $C := 0$

Enumerate sequentially all sequences $s \in S = \{0,1\}^d$

For each such sequence s:

If $\forall j \leq k [\sum_{i \leq d} s(i)x_{ji} = R_j \vee \sum_{i \leq d} s(i)x_{ji} = R_j - 1]$ then $C \longleftarrow C + 1$.

We note the length of sequences $s \in S$ is $d = r - k$, hence the brute force procedure has to enumerate 2^{r-k} members of S. Furthermore, each such sequence $s \in S$ is tested twice against all of the constraints $x_{1i}, x_{2i}, \ldots, x_{ki}$ for $i \leq d$, resulting in the claimed time complexity of $\mathcal{O}(2k2^{r-k})$.

To see the algorithm is correct, we give a proof that considers when the counter is incremented. Suppose for all $j \leq k$ some $s \in S$ is not a solution to either $\sum_{i \leq d} s(i)x_{ji} = R_j$ or $\sum_{i \leq d} s(i)x_{ji} = R_j - 1$. In this case, the counter is not incremented and we claim s does not induce a solution to the 1-3-SAT$^+$ formula φ. For in this case s is not a 0/1 solution to the system $\mathsf{Sys}(\varphi)$ and hence by Corollary 5.1 cannot be a satisfying solution to φ. In effect, the counter is not incremented as we have not seen an additional satisfying solution.

Now suppose for all $j \leq k$ some $s \in S$ is a solution to either $\sum_{i \leq d} s(i)x_{ji} = R_j$ or $\sum_{i \leq d} s(i)x_{ji} = R_j - 1$. In this case, the counter is incremented and we claim s is indeed a solution to the 1-3-SAT$^+$ formula φ.

For if s is a solution to all jth rows constraint $\sum_{i \leq d} s(i)x_{ji} = R_j$ then s satisfies the constraint $x_{j1} + x_{j2} + \cdots + x_{jd} = R_j$ giving the satisfying assignment $a(p) = \bot$ for all variables p corresponding to variables in the diagonal matrix, and $a(p) = \bot$ for variables corresponding to column i for which $s(i) = 0$, and $a(p) = \top$ for variables corresponding to column i for which $s(i) = 1$.

Similarly, if s is a solution to all jth rows constraint $\sum_{i \leq d} s(i)x_{ji} = R_j - 1$ then s satisfies the constraint $1 + x_{j1} + x_{j2} + \cdots + x_{jd} = R_j$ giving the satisfying assignment $a(p) = \top$ if p corresponds to the diagonal variable (j,j), $a(p) = \bot$ for all variables p corresponding to all other variables in the diagonal matrix, and $a(p) = \bot$ for variables corresponding to column i for which $s(i) = 0$, and $a(p) = \top$ for variables corresponding to column i for which $s(i) = 1$. □

Corollary 5.2. 1-3-SAT$^+ \leq_{poly}$ 0-1-IP$^=$.

Proof. By the pre-processing of the problem instance using Gaussian Elimination, shown above, 1-3-SAT$^+$ is reduced in polynomial time to 0-1-IP$^=$. □

Theorem 5.1. #1-3-SAT$^+\in \mathcal{O}(\eta 2^{\bar{\eta}+1})$ *for formula rank and nullity η and $\bar{\eta}$.*

Proof. There are η-many equations to satisfy by any assignment, and there are $\bar{\eta}$-many variables to search through exhaustively in order to solve the 0-1-IP$^=$ problem, which in turn solves the 1-3-SAT$^+$ problem. □

Corollary 5.3. #1-3-SAT$^+ \in \mathcal{O}(2\kappa r 2^{(1-\kappa)r})$ *for any instance $\varphi \in \Phi(r,k)$ and $\kappa = k/r$.*

6. Implications for positive 1-in-3 SAT

In [15] it is shown that 1-3-SAT can be solved in time $\mathcal{O}(2^{|V|/2})$ and space $\mathcal{O}(2^{|V|/4})$ through the 4-table method. The same result entails the same upper bounds for the 0-1-IP$^=$ problem.

Theorem 6.1 ([15])**.** *#1-3-SAT can be solved in time $\mathcal{O}(2^{|V|/2})$ and space $\mathcal{O}(2^{|V|/4})$.*

Proof sketch:. Split the problem instance in two partitions of roughly $|V|/2$ variables, solve each partition by exhaustive search and combine the two sets of solutions to obtain the solutions to the original problem. □

Theorem 6.2 ([15])**.** *#0-1-IP$^=$ can be solved in time $\mathcal{O}(2|C|2^{|V|/2})$ and space $\mathcal{O}(2^{|V|/4})$.*

Proof sketch:. Each constraint of 0-1-IP$^=$ can be seen as a subset-sum problem, hence solvable by the 4-table method. □

Corollary 6.1. *#1-3-SAT$^+$ can be solved in time in time $\mathcal{O}(4/3|V|2^{3|V|/8})$ and space $\mathcal{O}(4/3|V|2^{3|V|/16})$.*

The proof idea is to split the problem instance into quarters as per the method of [15] and instead of solving each quarter using as resources $\mathcal{O}(2^{|V|/2})$ time and $\mathcal{O}(2^{|V|/4})$ space, solve the associated 0-1-IP$^=$ instance of at most $3|V|/8$ variables in $\mathcal{O}(4/3|V|2^{3|V|/8})$ time and in $\mathcal{O}(4/3|V|2^{3|V|/16})$ space.

In other words, we split the #1-3-SAT$^+$ instance in two sub-formulas with roughly $|V|/2$ variables each and we may apply our kernel method by splitting each #0-1-IP$^=$ constraint in half again and performing an exhaustive search on the halved instances, as per the 4-table method of [15].

It is essential to note that the #0-1-IP$^=$ instance can be viewed as multiple subset-sum instances that can be solved in a divide-and-conquer fashion.

7. The method of Substitution

For ease of analysis we shall suppose Gauss-Jordan Elimination above is replaced by the method of substitution. The algorithm is depicted below in Fig. 1. Let $n(c), m(c), s(c)$ be the lowest, middle and highest labeled variable in clause c. By definition, these values must be distinct. Assume the formula φ is sorted in ascending order of $n(c)$. Represent clause c in equation form as $n(c) = 1 - m(c) - s(c)$.

```
i ← k
while i ≥ 1 do
    j ← k − 1
    while j ≥ 1 do
        if c(j) = A_j − (∑_r C(r)B(r)) − c(i)B(i), and c(i) = A_i − (∑_t C(t)B(t))
        then
            c(j) = (A_j − A_i B(i)) − (∑_r C(r)B(r) + ∑_t C(t)B(t)))
        end if
        j ← j − 1
    end while
    i ← i − 1
end while
```

Figure 1: Substitution algorithm

After the substitution process is finished, each of the clauses is expressed in terms of independent variables, variables which cannot be expressed in terms of other variables. We denote by $|n(i)|$ the number of variables in constraint $c(i)$ induced by the substitution method, excluding the variable $n(i)$.

Remark 7.1. *The largest number of expansions determined by running substitution on the collection of clauses, is $|n(k)| = 2$, $|n(k-1)| = 3$, $|n(k-2)| = 5, \ldots$, $|n(k-i)| = Fib(i+3)$.*

Proof. We need to maximize the number of substitutions performed at each step. Hence, at first step we encounter two substitutions, at the second we encounter three substitutions, while at every subsequent step we must assume there exist two variables for which we can substitute in terms of previously found variables, which indicates that the formula for the Fibonacci expansion describes our process. □

Definition 7.1 (Representation). *The size of a representation for a given instance of 1-3-SAT$^+$ $\in \Phi(r,k)$ expressed by substitution as $n(1), n(2), \ldots, n(k)$ is given by the formula*

$$r \times \log(\sum_{i \leq k} |n(i)|)$$

Remark 7.2. *The size of the resulting representation associated to formulas treated by Remark 7.1 converges asymptotically to $r^2 \times \log(1.62)$.*

Proof. The bound is given by an analysis of the growth of the Fibonacci sequence. It is well known the rate of growth of the sequence converges approximately to 1.62^n. □

Remark 7.3. *Contrast the scenario in Remark 7.1, to the case in which there are no substitutions induced, i.e. $\varphi = \{\{p_{3i+1}, p_{3i+2}, p_{3i+3}\}, i \leq 1/3k\}$.*

Remark 7.4. *The size of the resulting representation associated to formulas treated by Remark 7.3 is $r \times \log(2/3r)$.*

Proof. In this case we have $2k$ independent variables, for a value of k of $1/3r$. □

Theorem 7.1. *Any 1-3-SAT$^+$ formula admits a representation with size S for*

$$r \times \log(2/3r) \leq S \leq r^2 \times \log(1.62)$$

Remark 7.5. *The size of any representation is bounded above by $r^{2-\epsilon}$ for*

$$\epsilon = \frac{0.52}{\log(r)}$$

Proof. $r^{2-\epsilon} = r^2 \times \log(1.62)$ implies $2 - \epsilon = 2 + \frac{\log \log(1.62)}{\log(r)}$ and therefore

$$\epsilon = \frac{0.52}{\log(r)}$$

□

8. Implications for Computational Complexity

Dell and Melkebeek [10] give a rigorous treatment of the concept of "sparsification". In their framework, an oracle communication protocol for a language L is a communication protocol between two players. The first player is given the input x and is only allowed to run in time polynomial in the length of x. The second player is computationally unbounded, without initial access to x. At the end of communication, the first player should be able to decide membership in L. The cost of the protocol is the length in bits of the communication from the first player to the second.

Therefore, if the first player is able to reduce, in polynomial time, the problem instance significantly, the cost of communicating the "kernel" to the second player would also decrease, hence providing us with a very natural formal account for the notion of sparsification.

Jansen and Pieterse in [11] state and give a procedure for any instance of Exact Satisfiability with unbounded clause length to be reduced to an equivalent instance of the same problem with only $|V|+1$ clauses, for number of variables $|V|$. Their argument uses Gaussian Elimination and in essence their method corresponds to the method presented above. The concern regarding the number of clauses in 1-3-SAT$^+$ can be addressed, as we have done above, by observing that for any instance C of 3-cnf-SAT, the chain of polynomial-time parsimonious reductions $C \to \hat{C} \to \bar{C}$, for $\hat{C}$ and $\bar{C}$ instances of 1-3-SAT and 1-3-SAT$^+$ respectively, implies that the variables of $\hat{C}$ and $\bar{C}$ outnumber the clauses.

What is also claimed in [11] is that, assuming coNP $\not\subseteq$ NP $\setminus$ poly, no polynomial time algorithm can in general transform an instance of Exact Satisfiability of $|V|$-many variables to a significantly smaller equivalent instance, i.e. an instance encoded using $\mathcal{O}(|V|^{2-\epsilon})$ for any $\epsilon > 0$.

We believe it is already transparent that, in fact, we have obtained a significantly smaller kernel for 1-3-SAT$^+$ above, i.e. transforming parsimoniously an instance of $|V|$ variables to a "compressed" instance of 0-1-IP$^=$ of at most $2/3|V|$ variables.

Definition 8.1 (Constraint Satisfaction Problem). *A csp is a triple (S, D, T) where*

- *S is a set of variables,*

- *D is the discrete domain the variables may range over, and*

- *T is a set of constraints.*

Every constraint $c \in T$ is of the form (t, R) where t is a subset of S and R is a relation on D. An evaluation of the variables is a function $v \colon S \to D$. An evaluation v satisfies a constraint (t, R) if the values assigned to elements of t by v satisfies relation R.

Remark 8.1. 3-cnf-SAT *is a csp, rewritten in csp form as*

$$S = V, D = \{\top, \bot\},$$

members of T are defined as the 3-cnf-SAT *clauses together with the standard boolean function that evaluates disjunctions, i.e. $R(a,b,c) = \top$ iff at least one of a,b,c is true.*

Remark 8.2. 1-3-SAT *is a csp, rewritten in csp form as*

$$S = V, D = \{\top, \bot\},$$

members of T are defined as the 1-3-SAT *clauses together with the function $R(a,b,c) = \top$ iff exactly one of a,b,c is true.*

Remark 8.3. 1-3-SAT$^+$ *is a csp, rewritten in csp form as*

$$S = V^+, D = \{\top, \bot\},$$

members of T are defined as the 1-3-SAT *clauses together with the function $R(a,b,c) = \top$ iff exactly one of a,b,c is true.*

Remark 8.4. *In what follows we switch between notations and write a csp in a more general form, with a problem (S, D, T) written as $L \subseteq \mathbb{N} \times \Sigma^*$, with instances (k, x) such that $k = |S|$ and x a string representation of D and T.*

Definition 8.2 (Kernelization). *Let L, M be two parameterized decision problems, i.e. $L, M \subseteq \mathbb{N} \times \Sigma^*$ for some finite alphabet Σ.*

A kernelization for the problem L parameterized by k is a polynomial-time reduction of an instance (k, x) to an instance (k', x') such that:

- *$(k, x) \in L$ if and only if $(k', x') \in M$,*

- *$k' \in \mathcal{O}(k)$, and*

- *$|x'| \in \mathcal{O}(|x|)$.*

In the extended case of referring to the counting versions $\#L, \#M$ we additionally require the kernelization to be parsimonious and we refer to it as a parsimonious kernelization.

Definition 8.3 (Encoding). *An encoding of a problem $L \subseteq \mathbb{N} \times \Sigma^*$ is a bijection $h\colon L \to \mathbb{N}$ such that for any $(k,x) \in \mathbb{N} \times \Sigma^*$ we have $h(k,x) \in \mathcal{O}(|x|)$.*

Definition 8.4. *A non-trivial kernel for* 3-cnf-SAT *is a kernelization of this problem transforming any instance $\varphi \in \Phi(r,k)$ to an instance $(f(r), g(k))$ of an arbitrary* NP-complete *csp M, such that $f(r) \in \mathcal{O}(r)$ and $g(k) \leq h(k,r)$ with $h(k,r) \in \mathcal{O}(r^{3-\epsilon})$ for an encoding h of φ and some $\epsilon > 0$.*

Remark 8.5 (Dell and Melkebeek [10]). 3-cnf-SAT *admits a trivial kernel $(f(r), g(k))$ with $g(k) \leq h(k,r)$ and $h(k,r) \in \mathcal{O}(r^3)$.*

Proof sketch:. Represent the problem instance as a $r \times k$ matrix M with the value $M(i,j) = 1$ iff variable i is in constraint j, otherwise $M(i,j) = 0$. There are in total r^3 possible constraints for a general problem. The hypothesis is then that the number of constraints in a general instance cannot be reduced significantly, i.e. as to reduce a general instance to $g(k,r) \in \mathcal{O}(r^{3-\epsilon})$ for some positive ϵ. It is shown in [10] that if this hypothesis fails, then coNP $\subseteq$ NP \ poly and the Polynomial Hierarchy collapses to its third level. $\square$

Lemma 8.1 (Dell and Melkebeek [10]). *If* 3-cnf-SAT *admits a non-trivial kernel, then* coNP $\subseteq$ NP \ poly.

Definition 8.5. *A non-trivial kernel for* 1-3-SAT *is a kernelization of this problem transforming any instance $\varphi \in \Phi(r,k)$ to an equivalent instance $(f(r), g(k))$ of an arbitrary* NP-complete *csp M, such that $f(r) \in \mathcal{O}(r)$ and $g(k) \leq h(k,r)$ with $h(k,r) \in \mathcal{O}(r^{2-\epsilon})$ for an encoding h of φ and some $\epsilon > 0$.*

Remark 8.6 (Jansen and Pieterse [11]). 1-3-SAT *admits a kernel* $(f(r), g(k))$ *with* $g(k) \leq h(k,r)$ *and* $h(k,r) \in \mathcal{O}(r^2)$.

Proof sketch:. Gaussian Elimination may be used to reduce a general instance $\varphi(r, k)$ to an instance of $g(k) = r+1$ clauses. By a similar matrix representation as in Remark 8.5 we obtain that $g(k) \in \mathcal{O}(r^2)$. □

The following result is stated in [11].

Lemma 8.2 (Jansen and Pieterse [11]). *If* 1-3-SAT *admits a non-trivial kernel, then* coNP $\subseteq$ NP $\setminus$ poly.

Lemma 8.3. *If* 1-3-SAT$^+$ *admits a non-trivial kernel, then* 1-3-SAT *admits a non-trivial kernel.*

Proof. Let $\varphi \in \Phi(r, k)$ be an instance of 1-3-SAT. By Schaeffer's results it follows φ can be parsimoniously polynomial-time reduced to a 1-3-SAT$^+$ formula $\bar{\varphi} \in \Phi(r', k')$ with $r' = r + 4k$ and $k' = 3k$.

Assuming 1-3-SAT$^+$ admits a non-trivial kernel, this implies 1-3-SAT admits a non-trivial kernel, and therefore through Lemma 8.1 coNP $\subseteq$ NP $\setminus$ poly.

To spell this out, suppose we have non-trivial kernel $(f(r'), g(k'))$ for the problem 1-3-SAT, with $g(k') \leq h(k', r')$ and $h(k', r') \in \mathcal{O}(r'^{2-\epsilon})$. We observe using the reduction from 1-3-SAT, $f(r+4k) \leq f(r) + 4f(k) \leq 5f(r)$ and therefore $f(r') \in \mathcal{O}(r)$ and, we obtain via the reduction the existence of a non-trivial kernel for 1-3-SAT, that is $g(3k) \leq 3g(k) \leq 3h(k,r)$ with $h(k,r) \in \mathcal{O}(r^{2-\epsilon})$. □

Essentially the following result is a restatement of Corollary 5.3.

Theorem 8.1. 1-3-SAT$^+$ *admits a non-trivial kernel.*

Proof. Follows from Lemma 5.3 by observing the following strategy: the first player preprocesses the input in polynomial time using Gaussian Elimination and passes the input to the second player which makes use of its unbounded resources to provide a solution to this kernel.

It remains to be inferred that the cost of this computation is bounded non-trivially, i.e. $h(k,r) \in \mathcal{O}(r^{2-\epsilon})$ for $\epsilon > 0$. Follows from Lemma 5.3, for the instance of 0-1-IP$^=$ to which we reduce has at most $f(r') \leq 2/3r$ variables and at most $g(k') \leq r$ clauses. By the same argument as in Remark 8.5, we are able to store the resulting instance of 0-1-IP$^=$ in a $(2/3r + 1) \times r$ matrix M with polynomial-bounded entries, such that $M(i,j) = d$ iff d is the coefficient of variable i in constraint j, to which we add the result column.

From Remark 7.5 we obtain indeed that the bit representation of this kernel is indeed $r^{2-\epsilon}$ for some non-negative ϵ. Since Gaussian Elimination and Substitution are equivalent methods, this statement is correct. □

Corollary 8.1. *Every* 1-3-SAT$^+$ *instance* $\varphi \in \Phi(r,k)$ *admits a non-trivial kernel* $(f(r), g(k))$, *such that* $f(r) \in \mathcal{O}(r)$, $g(k) \leq h(k,r)$ *and* $h(k,r) \in \mathcal{O}(r^{2-\epsilon})$ *with* $\epsilon = \log(2/3)/\log(r)$.

Proof. We have established in Theorem 8.1 the kernel found in Lemma 5.3 may be stored in a matrix M with polynomial-bounded entries, such that $M(i,j) = d$ iff d is the coefficient of variable i in constraint j, and we have established further that $i \leq 2/3|V| + 1$, $j \leq |V|$ and $d \in \mathcal{O}(|V|^3)$. It follows that the kernel size is at most $2/3\bar{C} \times |V|^5$ for some constant $\bar{C}$, which gives a size in bits for the kernel of $\log(C) + 5\log(|V|)$ for $C = 2/3\bar{C}$.

□

Corollary 8.2. $\text{coNP} \subseteq \text{NP} \setminus \text{P}$

Proof. Follows from Lemma 8.3, Theorem 8.1 and Lemma 8.2. □

9. Conclusion

Our contribution is in pointing out the relevance of Gaussian Elimination in dealing with certain types of constraint satisfaction problems, such as the ones that presuppose a type of exclusivity between constraints.

These problems may be encoded into linear systems of equations which can be solved using the method of Gaussian Elimination, or the equivalent method of Substitution.

The challenge we faced was a rigorous analysis of the fully defined linear systems, together with a type of brute-force approach in solving these systems.

Another important detail we had to analyze rigorously was the issue of growth in size of representation of the matrix in question.

The most important question in Theoretical Computer Science remains open.

Acknowledgments

Foremost thanks are due to Igor Potapov for his support and benevolence.

Most of the ideas presented have crystallized while the author was studying with Rod Downey at Victoria University of Wellington, in the New Zealand winter of 2010.

I am very much indebted to Noam Greenberg, Dillon Mayhew, Cristian Calude, Rob Goldblatt, Max Cresswell, Ed Mares, Mark Reynolds and Tim French for supervising various projects in which I was involved.

Special acknowledgments are given to Reino Niskanen for many useful comments and for proof reading an initial version of this manuscript.

References

[1] S. A. Cook, The complexity of theorem-proving procedures, in: Proceedings of the third annual ACM symposium on Theory of computing, ACM, 1971, pp. 151–158.

[2] R. M. Karp, Reducibility among combinatorial problems, in: Complexity of computer computations, Springer, 1972, pp. 85–103.

[3] L. A. Levin, Universal sequential search problems, Problemy Peredachi Informatsii 9 (3) (1973) 115–116.

[4] T. J. Schaefer, The complexity of satisfiability problems, in: Proceedings of the tenth annual ACM symposium on Theory of computing, ACM, 1978, pp. 216–226.

[5] V. Dahllöf, P. Jonsson, R. Beigel, Algorithms for four variants of the exact satisfiability problem, Theoretical Computer Science 320 (2-3) (2004) 373–394.

[6] A. Björklund, T. Husfeldt, Exact algorithms for exact satisfiability and number of perfect matchings, Algorithmica 52 (2) (2008) 226–249.

[7] M. Soos, Enhanced gaussian elimination in dpll-based sat solvers., in: POS@ SAT, 2010, pp. 2–14.

[8] M. Wahlström, Abusing the tutte matrix: An algebraic instance compression for the k-set-cycle problem, arXiv preprint arXiv:1301.1517.

[9] A. C. Giannopoulou, D. Lokshtanov, S. Saurabh, O. Suchy, Tree deletion set has a polynomial kernel but no opt^o(1) approximation, SIAM Journal on Discrete Mathematics 30 (3) (2016) 1371–1384.

[10] H. Dell, D. Van Melkebeek, Satisfiability allows no nontrivial sparsification unless the polynomial-time hierarchy collapses, Journal of the ACM (JACM) 61 (4) (2014) 23.

[11] B. M. Jansen, A. Pieterse, Optimal sparsification for some binary csps using low-degree polynomials, arXiv preprint arXiv:1606.03233.

[12] B. M. Jansen, A. Pieterse, Sparsification upper and lower bounds for graph problems and not-all-equal sat, Algorithmica 79 (1) (2017) 3–28.

[13] J. Ding, A. Sly, N. Sun, Proof of the satisfiability conjecture for large k, in: Proceedings of the forty-seventh annual ACM symposium on Theory of computing, ACM, 2015, pp. 59–68.

[14] R. G. Downey, M. R. Fellows, Fundamentals of parameterized complexity, Vol. 201, Springer, 2016.

[15] R. Schroeppel, A. Shamir, A t=o($2^n/2$), s=o($2^n/4$) algorithm for certain np-complete problems, SIAM journal on Computing 10 (3) (1981) 456–464.

[16] L. G. Valiant, The complexity of computing the permanent, Theoretical computer science 8 (2) (1979) 189–201.

[17] M. R. Garey, D. S. Johnson, Computers and intractability, W.H. Freeman, New York, 1979.

The Polynomial Hierarchy Collapses.
A Fresh Perspective.

Valentin Bura[a]

[a]*Department of Computer Science, University of Liverpool*

Abstract

In a previous article we prove the Polynomial Hierarchy collapses by making use of a method based essentially on Gaussian Elimination. In this paper we replace completely our use of Gaussian Elimination and formulate an equivalent approach based on the method of Substitution. We give the usual argument that a non-trivial kernel for Exact Satisfiability may be found. Our proof shows the structure formerly known as the Polynomial Hierarchy collapses to the level above $P = NP$. That is, we show that $\mathtt{coNP} \subseteq \mathtt{NP} \setminus \mathtt{P}$.

Keywords: Computational Complexity, Boolean Satisfiability, Kernelization

---

1. Introduction

One-in-three satisfiability was first studied in the late seventies as an elaboration relating to Schaefer's Dichotomy Theorem [1].

It is proved in [1], using certain assumptions, that boolean satisfiability problems are either in `P` or they are `NP-complete`.

In [2] we give a method based on Gaussian Elimination and we use this method to construct a non-trivial kernel for `1-3-SAT`$^+$. The implication is that

$$\mathtt{coNP} \subseteq \mathtt{NP} \setminus \mathtt{P}$$

Boolean Satisfiability `SAT` and its restrictions `cnf-SAT`, `k-cnf-SAT`, `3-cnf-SAT` are `NP-complete` as shown in [3, 4, 5].

The 1-3-SAT problem [1] is that, given a collection of triples over some variables, to determine whether there exists a truth assignment to the variables so that each triple contains exactly one true literal.

1-3-SAT is well-known to be complete for the class NP while a parsimonious reduction from 1-3-SAT also shows 1-3-SAT$^+$ to be complete for NP.

It is widely believed counting to be harder than the corresponding decision problem, since counting sat-assignments of a formula in 2-CNF is complete for #P, while the decision problem is known to be in P [6].

We mention Sinosuke Toda showed in [7] counting to be as hard as the Polynomial Hierarchy. Our results here imply counting to be just as hard as NP, this being the essential implication of our result of [2].

In [8] the author uses a similar method to ours for handling xor types of constraints. Other recent examples of Gaussian elimination used in exact algorithms or kernelization may be indeed found in the literature [9, 10].

Hence the idea that constraints of the type implying this type of exclusivity can be formulated in terms of equations, and therefore processed using Gaussian Elimination or an equivalent method, is not new and the intuition behind it is very straightforward.

Our method shows that positive instances of 1-in-3 SAT may be reduced to significantly smaller instances of I.P. That is, any such instance with $|V|$ variables and $|C|$ clauses can be poly-time reduced to an instance of 0/1 Integer Programming with equality only, of size at most $2/3|V|$ and $|C|$ clauses.

The extremely influential papers of Dell and Van Melkebeek [11], and of Jansen and Pieterse [12, 13] show that, under the assumption that coNP $\not\subseteq$ NP $\setminus$ P, there cannot exist a "significantly small" kernel for various problems, of which exact 1-3-SAT is one. We use these results directly in our current approach, as we have done in [2].

Our preprocessing induces a certain type of "order" on the variables, such that some of the non-satisfying assignments can be omitted by the solution search.

We define Substitution and show this method alone suffices for identifying a non-trivial kernel for `1-3-SAT`$^+$.

We then proceed to define formally the notion of a non-trivial kernel. For this, we define the problems considered as Constraint Satisfaction Problems.

We conclude by showing that the method presented here, giving a non-trivial kernel for `1-3-SAT`$^+$, implies the existence of a non-trivial kernel for `1-3-SAT`.

2. Notation

We denote boolean variables by $p_1, p_2, \ldots, p_i, \ldots$ Denote the true and false constants by $\top$ and $\bot$ respectively. For any SAT formula φ, write $\Sigma(\varphi)$ if φ is satisfiable and write $\bar{\Sigma}(\varphi)$ otherwise. Reserve the notation $a(p)$ for a truth assignment to the variable p.

We write $\Phi(r, k)$ for the set of formulas in 3-CNF with r variables and k unique clauses. We also write $\varphi(V, C)$ to specify concretely such a formula, where V, C shall denote the sets of variables and clauses of φ. We write $\kappa(\varphi) = \frac{k}{r}$.

We will make use of the following properties of a given map f:

subadditivity: $f(A + B) \leq f(A) + f(B)$

scalability: $f(cA) = cf(A)$ for constant c.

For a given tuple $s = (s_1, s_2, \ldots, s_n)$ we let $s(m)$ denote the element s_m. Finally, for given linear constraints $L: \sum_{i \leq n} d_i x_i = R$ for some n and $x_i \in \{0, 1\}$, denote by $coef(x_i)$ the value d_i.

Let $L[x/L']$ be the result of substituting uniformly the expression of constraint L' for variable x in constraint L, to be performed in restricted circumstances.

3. Exact Satisfiability

Definition 3.1 (1-3-SAT). *1-3-SAT is defined as determining whether $\varphi \in \Phi(r,k)$ is satisfiable, where the formula is defined as*

$$\varphi = \{\{p_1^1, p_2^1, p_3^1\}, \{p_1^2, p_2^2, p_3^2\}, \ldots, \{p_1^k, p_2^k, p_3^k\}\}$$

such that $p_1^i, p_2^i, p_3^i \in V = \{p_1, \neg p_1, p_2, \neg p_2, \ldots, p_r, \neg p_r\} \cup \{\bot\}$

For any clause exactly one of the literals is allowed to be true in an assignment.

No clause is allowed to contain repeated literals or a literal and its negation.

Every variable in V appears in at least one clause.

In the restricted case that $p_1^i, p_2^i, p_3^i \in V^+ = \{p_1, p_2, \ldots, p_r\} \cup \{\bot\}$ for $1 \leq i \leq r$ we denote the problem as 1-3-SAT$^+$.

Example 3.1. *The 1-3-SAT$^+$ formula $\varphi = \{\{p_1, p_2, p_3\}, \{p_2, p_3, p_4\}\}$ is satisfiable by the assignment $a(p_2) = \top$ and $a(p_j) = \bot$ for $j = 1, 3, 4$. The 1-3-SAT$^+$ formula $\varphi = \{\{p_1, p_2, p_3\}, \{p_2, p_3, p_4\}, \{p_1, p_2, p_4\}, \{p_1, p_3, p_4\}\}$ is not satisfiable.*

Lemma 3.1 (Bura [2]). *Up to uniqueness of clauses and variable naming the set $\Phi(r, r/3)$ determines one 1-3-SAT$^+$ formula and this formula is trivially satisfiable.*

Remark 3.1 (Bura [2]). *For 1-3-SAT$^+$, the sets $\Phi(r, k)$ for $k < r/3$ are empty.*

Schaefer [1] gives a polynomial time parsimonious reduction from 3-cnf-SAT to 1-3-SAT hence showing that 1-3-SAT and its counting version #1-3-SAT are NP-complete and respectively #P-complete.

Proposition 3.1 (Schaefer, [1]). 1-3-SAT *is* NP-complete.

Proposition 3.2 (Garey and Johnson[14]). 1-3-SAT$^+$ *is* NP-complete.

Remark 3.2 (Bura [2]). *In virtue of Theorem 3.1 and Theorem 3.2 we restrict our study to instances of* 1-3-SAT$^+$ $\varphi \in \Phi(r, k)$ *with* $r \geq k$.

For if an instance of 3-cnf-SAT $\bar{\varphi} \in \Phi(r', k')$ *is reduced to an instance of* 1-3-SAT $\hat{\varphi} \in \Phi(r'', k'')$ *then our reduction entails* $r'' = r' + 4k'$ *and* $k'' = 3k'$.

We analyze the further reduction to the instance of 1-3-SAT$^+$ $\varphi \in \Phi(r, k)$. *Let* C, C', C'', C''' *be the collections of clauses in* $\hat{\varphi}$ *containing, no negation, one negation, two negations and three negations respectively.*

Our reduction implies $r = r'' + |C'| + 2|C''| + 3|C'''|$ *and* $k = |C| + 2|C'| + 3|C''| + 4|C'''| = k'' + |C'| + 2|C''| + 3|C'''|$. *Then,* $r - k = r'' + |C'| + 2|C''| + 3|C'''| - k'' - |C'| - 2|C''| - 3|C'''| = r'' - k'' = r' + 4k' - 3k' = r' + k' > 0$.

4. Rank of a Formula

Rank and nullity are used as measures of independence, and of dependence, for members of a given set.

Definition 4.1. *A rank function R obeys the following*

1. $R(\emptyset) = 0$,

2. $R(A \cup B) \leq R(A) + R(B)$,

3. $R(A) \leq R(A \cup \{a\}) \leq R(A) + 1$.

Definition 4.2 (rank and nullity). *For a* 1-3-SAT$^+$ *formula $\varphi(V, C)$ define the system of linear equations* $\text{Sys}(\varphi)$ *as follows:*

for any clause $\{p, p', p''\} \in C$ add to $\text{Sys}(\varphi)$ *equation $\bar{p} + \bar{p}' + \bar{p}'' = 1$;*

Define the rank and nullity of φ as $\eta(\varphi) = R(\text{Sys}(\varphi))$ and $\bar{\eta}(\varphi) = |V| - \eta(\varphi)$. If formula is clear from context, we also use the shorthand η and $\bar{\eta}$.

Remark 4.1 (Bura [2]). *η is a rank function with respect to sets of* 1-3-SAT *triples.*

Lemma 4.1. *For any* 1-3-SAT$^+$ *instance φ transformed into a linear system* $\text{Sys}(\varphi)$ *one observes the following:*

$\bar{p} + \bar{p}' + \bar{p}'' = 1$ has a solution $S \subset \{0,1\}^3$ if and only if exactly one of $\bar{p}, \bar{p}', \bar{p}''$ is equal to 1 and the other two are equal to 0.

Proposition 4.1. *For any formula $\varphi \in \Phi(r, k)$ we have $\Sigma(\varphi)$ if and only if* $\text{Sys}(\varphi)$ *has at least one solution over $\{0,1\}^r$.*

Corollary 4.1. *A formula $\varphi \in \Phi(r, k)$ has as many satisfiability assignments as the number of solutions of* $\text{Sys}(\varphi)$ *over $\{0,1\}^r$.*

We define the binary integer programming problem with equality here and infer briefly that 1-3-SAT$^+$ is reducible to a "smaller" instance of this problem.

Definition 4.3 (0, 1-integer programming with equality). *The* 0-1-IP$^=$ *problem is defined as follows. Given a family of finite tuples $s_1, s_2, \ldots, s_k$ with each $s_i \in \mathbb{Q}^S$ for some fixed $S \in \mathbb{N}$, and given a sequence $q_1, q_2, \ldots, q_k \in \mathbb{Q}$, decide whether there exists a tuple $T \in \{0,1\}^S$ such that*

$$\sum_{j=1}^{S} s_i(j)T(j) = q_i \text{ for each } i \in \{1,2,\ldots,k\}$$

The following bound is obtained through applying an exhaustive search.

Remark 4.2 (Bura [2]). $\texttt{0-1-IP}^= \in \mathcal{O}(k2^S)$ where k is the number of $\texttt{0-1-IP}^=$ tuples and S is the size of the tuples.

An observation on the rank of a formula, which cannot exceed the number of clauses.

Lemma 4.2 (Bura [2]). Let $\varphi \in \Phi(r,k)$ be a $\texttt{1-3-SAT}^+$ formula, then $\eta(\varphi) \leq k$ and $\bar{\eta}(\varphi) \geq r - k$.

We give a bound for the case of the formula being full-rank.

Lemma 4.3 (Bura [2]). Consider a $\texttt{1-3-SAT}^+$ formula φ and suppose $\eta(\varphi) = k$ and $\bar{\eta}(\varphi) = r - k$. The satisfiability of φ is decidable in $\mathcal{O}(2k2^{r-k})$.

We mention the polynomial-time reduction to Integer Programming with Equality only.

Corollary 4.2 (Bura [2]). $\texttt{1-3-SAT}^+ \leq_{poly} \texttt{0-1-IP}^=$.

Hence, we give the worst-time scenario for a brute-force approach, applied to I.P.

Proposition 4.2 (Bura [2]). $\#\texttt{1-3-SAT}^+ \in \mathcal{O}(\eta 2^{\bar{\eta}+1})$ for formula rank and nullity η and $\bar{\eta}$.

And an interesting variation on problem hardness, with modifications on the clauses-to-variables ratio.

Corollary 4.3 (Bura [2]). $\#\texttt{1-3-SAT}^+ \in \mathcal{O}(2\kappa r 2^{(1-\kappa)r})$ for any instance $\varphi \in \Phi(r,k)$ and $\kappa = k/r$.

5. The Method of Substitution

The substitution algorithm is depicted below in Fig. 1. We give a brief textual explanation of the algorithm below.

Pre-processing phase

1. Let $n(c), m(c), s(c)$ be the lowest, middle and highest labeled variable in clause c. These values are distinct.

2. Represent clause c in normal form as $n(c) = 1 - m(c) - s(c)$.

3. Sort the formula φ in ascending order of $n(c)$.

Substitution phase

1. Initialize $i \leftarrow |C|$,

2. For each c_i: $n(i) = a(i) - m(i) - s(i) = a(i) - M(i)$,

3. Initialize $j \leftarrow |C|$,

4. For each clause $c_j \in \varphi$ with $j \neq i$ with c_j: $n(j) = a(j) - m(j) - s(j)$ such that $n(j)$ is found in the variables of $m(i)$, or $n(j)$ is found in the variables of $s(i)$, do

5. Perform the substitution $c_i \leftarrow c_i[n(j)/n(i)]$, and normalize the result.

6. Decrement variable j. Continue step 4.

7. Decrement variable i. Continue step 2.

```
i ← k

while i ≥ 1 do

    j ← k

    while j ≥ 1 do

        if j ≠ i then

            if c(j): n(j) = a(j) − (∑_{x<r} d(x) + n(i)), and c(i): n(i) = a(i) − (∑_{t<r} d(t))
            then

                c(j) ← n(j) = (a(j) − a(i)) − (∑_{x<r} d(x) + ∑_{t<r} d(t))

            end if

        end if

        j ← j − 1

    end while

    i ← i − 1

end while
```

Figure 1: Substitution algorithm

We remark an essentially cubic halting time on the substitution algorithm, which intuitively corresponds to the cubic halting time of Gaussian Elimination, an equivalent method.

Remark 5.1. *Substitution halts in time $\mathcal{O}(k^2 \times r)$ for any formula $\varphi \in \Phi(k,r)$.*

Denote by $\sigma(\varphi)$ or by σ, when clear from context, the structure thus obtained, denote by $\eta(\sigma)$ and $\bar{\eta}(\sigma)$ the rank and nullity thus induced, and denote by $N(\sigma)$ and $\bar{N}(\sigma)$ the sets of independent, and dependent variables generated through our process.

We remark the operator σ is idempotent.

Remark 5.2. $\sigma(\sigma(\varphi)) = \sigma(\varphi)$.

Proof. Each clause $c(j)$ is read, and each read clause is compared with every other clause $c(i)$, in search for a common variable $n(i)$, if this variable is found, a replacement is performed on $c(j)$.

Suppose there exists a clause c such that $L = \sigma(\varphi)(c) \neq \sigma(\sigma(\varphi))(c) = L'$.

Consider the case $L \setminus L' \neq \emptyset$. Let variable v be in this set difference. It cannot be the case that $v = n(c)$ since this means the procedure missed a mandatory substitution of $n(c)$, which the second iteration picked up.

Therefore $v \neq n(c)$. In this case, v is a result of a single substitution, or of a chain of substitutions ending with c. An induction on this chain of substitutions, shows the procedure missed a mandatory substitution of an n-variable, which the second iteration picked up.

Consider the case $L' \setminus L \neq \emptyset$. Let variable v' be in this set difference. It cannot be the case that $v' = n(c)$ since this means the procedure introduced a new substitution of $n(c)$, which the first iteration missed.

Therefore $v \neq n(c)$. In this case, v is a result of a single substitution, or of a chain of substitutions ending with c. An induction on this chain of substitutions, shows the first iteration of the procedure missed a mandatory substitution of an n-variable which the second iteration picked up. $\square$

As a consequence, any set of formulas is closed under substitution.

Remark 5.3. $\sigma[\sigma[\Phi]] = \sigma[\Phi]$.

6. An example

Consider the `1-3-SAT` formula

$$\varphi = \{\{p_1, p_2, p_3\}, \{p_4, p_5, p_6\}, \{p_2, p_5, p_6\}, \{p_1, p_2, p5\}\}$$

We outline the meaning of the rows and columns within our tabular format.

c_i:	$n(i)$	$m(i)$	$c(i)$

$C(i)$:	$n(i)$	$a(i)$	$m(i)$	$c(i)$

The formula φ is represented in tabular format. Sort according to $n(i)$.

c_1:	1	2	3
c_2:	4	5	6
c_3:	2	5	6
c_4:	1	2	5

c_1:	1	2	3
c_4:	1	2	5
c_3:	2	5	6
c_2:	4	5	6

The formula is encoded as below. Use a tabular data structure for the algorithm, initialized to empty.

1.

$C(1)$:	∅	∅	∅	∅
$C(4)$:	∅	∅	∅	∅
$C(3)$:	∅	∅	∅	∅
$C(2)$:	∅	∅	∅	∅

2.

$C(1)$:	1	0	2	3
$C(4)$:	1	0	2	5
$C(3)$:	2	0	5	6
$C(2)$:	4	0	5	6

Substitution phase. Operate on the data structure.

3.

$C(1)$:	1	1	2	3
$C(4)$:	1	1−1	5,6	5
$C(3)$:	2	1	5	6
$C(2)$:	4	1	5	6

4.

$C(1)$:	1	1−1	5,6	3
$C(4)$:	1	1−1	5,6	5
$C(3)$:	2	1	5	6
$C(2)$:	4	1	5	6

Obtain the following partial result.

5.

$C(1)$:	1	0	5,6	3
$C(4)$:	1	0	5,6	5
$C(3)$:	2	1	5	6
$C(2)$:	4	1	5	6

Rearrange the tabular structure.

C_1:	p_1	=	0	−	p_3, p_5, p_6
C_4:	p_1	=	1	−	p_5, p_5, p_6
C_3:	p_2	=	1	−	p_5, p_6
C_2:	p_4	=	1	−	p_5, p_6

Note the independent variables are $\{p_1, p_2, p_4\}$ hence the rank and nullity of the formula are $\eta(\varphi) = 3$ and $\bar{\eta}(\varphi) = 3$. A Brute-Force Search on the set $\{p_3, p_5, p_6\}$ of dependent variables yields the desired result to the 1-3-SAT$^+$ formula.

After the substitution process is finished, each of the clauses is expressed in terms of independent variables, variables which cannot be expressed in terms of other variables. We denote by $|n(i)|$ the number of variables in constraint $c(i)$ induced by the substitution method, excluding the variable $n(i)$.

7. Algorithm Analysis

We maximize the number of substitutions performed at each step. Hence, at first step we encounter two substitutions, at the second we encounter three substitutions, while at every subsequent step we must assume there exist two variables for which we can substitute in terms of previously found variables, which indicates that the formula for the Fibonacci expansion describes our process.

Remark 7.1 (Bura [2]). *The largest number of expansions determined by running substitution on the collection of clauses, is $|n(k)| = 2$, $|n(k-1)| = 3$, $|n(k-2)| = 5, \ldots, |n(k-i)| = Fib(i+3)$.*

Definition 7.1 (Representation). *The size of a representation for a given instance of* 1-3-SAT$^+$ $\in \Phi(r,k)$ *expressed by substitution as $n(1), n(2), \ldots, n(k)$ is given by the formula*
$$r \times \log(\sum_{i \leq k} |n(i)|)$$

Remark 7.2. *The size of the resulting representation associated to formulas treated by Remark 7.1 converges asymptotically to $r^2 \times \log(1.62)$.*

Proof. The bound is given by an analysis of the growth of the Fibonacci sequence. It is well known the rate of growth of the sequence converges approximately to 1.62^n. □

Remark 7.3. *Contrast the scenario in Remark 7.1, to the case in which there are no substitutions induced, i.e.* $\varphi = \{\{p_{3i+1}, p_{3i+2}, p_{3i+3}\}, i \leq 1/3k\}$.

Remark 7.4. *The size of the resulting representation associated to formulas treated by Remark 7.3 is $r \times \log(2/3r)$.*

Proof. In this case we have $2k$ independent variables, for a value of k of $1/3r$. □

Theorem 7.1. *Any* 1-3-SAT$^+$ *formula admits a representation with size S for*

$$r \times \log(2/3r) \leq S \leq r^2 \times \log(1.62)$$

Remark 7.5. *The size of any representation is bounded above by $r^{2-\epsilon}$ for*

$$\epsilon = \frac{0.52}{\log(r)}$$

Proof. $r^{2-\epsilon} = r^2 \times \log(1.62)$ implies $2 - \epsilon = 2 + \frac{\log \log(1.62)}{\log(r)}$ and therefore

$$\epsilon = \frac{0.52}{\log(r)}$$

$\square$

8. Adequacy Proof

Proposition 8.1. *Let φ be a* 1-3-SAT$^+$ *formula and let $\sigma = \sigma(\varphi)$ be the resulting structure obtained by performing substitution on φ. Then, $\eta(\varphi) \leq \eta(\sigma)$ and $\bar{\eta}(\varphi) \geq \bar{\eta}(\sigma)$.*

Proof. It suffices to show that $\bar{\eta}(\varphi) \geq \bar{\eta}(\sigma)$.

Suppose for a contradiction this is not the case. We have that $\bar{\eta}(\varphi) < \bar{\eta}(\sigma)$.

That is, that the dependent variables of the system of equations exceed in number the dependent variables obtained through our substitution algorithm.

We let $\bar{\eta}(\varphi) = \bar{\eta}(\sigma) + K$. What this means is there exist variables $p_1, p_2, \cdots, p_K$ such that $p_i \in \bar{N}(\sigma) \setminus \bar{N}(\varphi)$ for $1 \leq i \leq K$.

Take any such variable in this list and perform another substitution such as to decrease K by one. The existence of the list $p_1, p_2, \cdots, p_K$ hence violates Remark 5.2. $\square$

9. Implications

Proposition 9.1 (Schroeppel and Shamir[15]). *#1-3-SAT can be solved in time $\mathcal{O}(2^{|V|/2})$ and space $\mathcal{O}(2^{|V|/4})$.*

Proposition 9.2 (Schroeppel and Shamir[15]). *#0-1-IP$^=$ can be solved in time $\mathcal{O}(2|C|2^{|V|/2})$ and space $\mathcal{O}(2^{|V|/4})$.*

Corollary 9.1. *#1-3-SAT$^+$ can be solved in time in time $\mathcal{O}(4/3|V|2^{3|V|/8})$ and space $\mathcal{O}(4/3|V|2^{3|V|/16})$.*

Dell and Melkebeek [11] give a rigorous treatment of the concept of "sparsification". In their framework, an oracle communication protocol for a language L is a communication protocol between two players.

The first player is given the input x and is only allowed to run in time polynomial in the length of x. The second player is computationally unbounded, without initial access to x. At the end of communication, the first player should be able to decide membership in L. The cost of the protocol is the length in bits of the communication from the first player to the second.

Therefore, if the first player is able to reduce, in polynomial time, the problem instance significantly, the cost of communicating the "kernel" to the second player would also decrease, hence providing us with a very natural formal account for the notion of sparsification.

Jansen and Pieterse in [12] state and give a procedure for any instance of Exact Satisfiability with unbounded clause length to be reduced to an equivalent instance of the same problem with only $|V|+1$ clauses, for number of variables $|V|$.

The concern regarding the number of clauses in 1-3-SAT⁺ can be addressed, as we have done above. We observe that for any instance C of 3-cnf-SAT, the chain of polynomial-time parsimonious reductions $C \to \hat{C} \to \bar{C}$, for $\hat{C}$ and $\bar{C}$ instances of 1-3-SAT and 1-3-SAT⁺ respectively, implies that the variables of $\hat{C}$ and $\bar{C}$ outnumber the clauses.

What is also claimed in [12] is that, assuming $\text{coNP} \not\subseteq \text{NP} \setminus \text{P}$, no polynomial time algorithm can in general transform an instance of Exact Satisfiability of $|V|$-many variables to a significantly smaller equivalent instance, i.e. an instance encoded using $\mathcal{O}(|V|^{2-\epsilon})$ for any $\epsilon > 0$.

We believe it is already transparent that, in fact, we have obtained a significantly smaller kernel for 1-3-SAT⁺ above, i.e. transforming parsimoniously an instance of $|V|$ variables to a "compressed" instance of 0-1-IP= of at most $2/3|V|$ variables.

Definition 9.1 (Constraint Satisfaction Problem). *A csp is a triple (S, D, T) where*

- *S is a set of variables,*

- *D is the discrete domain the variables may range over, and*

- *T is a set of constraints.*

Every constraint $c \in T$ is of the form (t, R) where t is a subset of S and R is a relation on D. An evaluation of the variables is a function $v \colon S \to D$. An evaluation v satisfies a constraint (t, R) if the values assigned to elements of t by v satisfies relation R.

Remark 9.1. *The following are constraint satisfaction problems:*

- 3-cnf-SAT

- 1-3-SAT

- 1-3-SAT$^+$

In what follows we switch between notations and write a csp in a more general form, with a problem (S, D, T) written as $L \subseteq \mathbb{N} \times \Sigma^*$, with instances (k, x) such that $k = |S|$ and x a string representation of D and T.

Definition 9.2 (Kernelization)**.** *Let L, M be two parameterized decision problems, i.e. $L, M \subseteq \mathbb{N} \times \Sigma^*$ for some finite alphabet Σ.*

A kernelization for the problem L parameterized by k is a polynomial time reduction of an instance (k, x) to an instance (k', x') such that:

- *$(k, x) \in L$ if and only if $(k', x') \in M$,*

- *$k' \in \mathcal{O}(k)$, and*

- *$|x'| \in \mathcal{O}(|x|)$.*

Definition 9.3 (Encoding)**.** *An encoding of a problem $L \subseteq \mathbb{N} \times \Sigma^*$ is a bijection $h \colon L \to \mathbb{N}$ such that for any $(k, x) \in \mathbb{N} \times \Sigma^*$ we have $h(k, x) \in \mathcal{O}(|x|)$.*

Definition 9.4. *A non-trivial kernel for 3-cnf-SAT is a kernelization of this problem transforming any instance $\varphi \in \Phi(r, k)$ to an instance $(f(r), g(k))$ of an arbitrary NP-complete csp M, such that $f(r) \in \mathcal{O}(r)$ and $g(k) \leq h(k, r)$ with $h(k, r) \in \mathcal{O}(r^{3-\epsilon})$ for an encoding h of φ and some $\epsilon > 0$.*

Remark 9.2 (Dell and Melkebeek [11]). `3-cnf-SAT` *admits a trivial kernel* $(f(r), g(k))$ *with* $g(k) \leq h(k,r)$ *and* $h(k,r) \in \mathcal{O}(r^3)$.

Lemma 9.1 (Dell and Melkebeek [11]). *If* `3-cnf-SAT` *admits a non-trivial kernel, then* `coNP` $\subseteq$ `NP` $\setminus$ `P`.

Definition 9.5. *A non-trivial kernel for* `1-3-SAT` *is a kernelization of this problem transforming any instance* $\varphi \in \Phi(r,k)$ *to an instance* $(f(r), g(k))$ *of an arbitrary* `NP-complete` *csp* M, *such that* $f(r) \in \mathcal{O}(r)$ *and* $g(k) \leq h(k,r)$ *with* $h(k,r) \in \mathcal{O}(r^{2-\epsilon})$ *for an encoding* h *of* φ *and some* $\epsilon > 0$.

Remark 9.3 (Jansen and Pieterse [12]). `1-3-SAT` *admits a kernel* $(f(r), g(k))$ *with* $g(k) \leq h(k,r)$ *and* $h(k,r) \in \mathcal{O}(r^2)$.

The following statement is given in [12]. The authors elaborate on the results of [11] to analyze combinatorial problems from the perspective of sparsification, and give several arguments that non-trivial kernels for such problems would entail a collapse of the Polynomial Hierarchy to the level above P = NP.

It is essential to note here that this line of reasoning was used by researchers studying sparsification with the intention of proving lower bounds on the existence of kernels, while the results presented by us are slightly more optimistic.

Lemma 9.2 (Jansen and Pieterse [12]). *If* `1-3-SAT` *admits a non-trivial kernel, then* `coNP` $\subseteq$ `NP` $\setminus$ `P`.

Lemma 9.3 (Bura [2]). *If* `1-3-SAT`$^+$ *admits a non-trivial kernel, then* `1-3-SAT` *admits a non-trivial kernel.*

Proof. Let $\varphi \in \Phi(r,k)$ be an instance of `1-3-SAT`. By Schaeffer's results it follows φ can be parsimoniously polynomial time reduced to a `1-3-SAT`$^+$ formula $\bar{\varphi} \in \Phi(r', k')$ with $r' = r + 4k$ and $k' = 3k$.

Assuming 1-3-SAT$^+$ admits a non-trivial kernel, this implies 1-3-SAT admits a non-trivial kernel, and therefore through Lemma 9.1 coNP $\subseteq$ NP $\setminus$ P.

To spell this out, suppose we have non-trivial kernel $(f(r'), g(k'))$ for the problem 1-3-SAT, with $g(k') \leq h(k', r')$ and $h(k', r') \in \mathcal{O}(r'^{2-\epsilon})$. We observe using the reduction from 1-3-SAT, $f(r + 4k) \leq f(r) + 4f(k) \leq 5f(r)$ and therefore $f(r') \in \mathcal{O}(r)$ and, we obtain via the reduction the existence of a non-trivial kernel for 1-3-SAT, that is $g(3k) \leq 3g(k) \leq 3h(k, r)$ with $h(k, r) \in \mathcal{O}(r^{2-\epsilon})$. $\square$

Essentially the following result is a restatement of Corollary 4.3.

Theorem 9.1 (Bura [2]). *1-3-SAT$^+$ admits a non-trivial kernel.*

Proof. Follows from Lemma 4.3. The first player preprocesses the input in polynomial time using Substitution, and passes the input to the second player which makes use of its unbounded resources to provide a solution to this kernel.

It remains to show the cost of this computation is bounded non-trivially, i.e. $h(k, r) \in \mathcal{O}(r^{2-\epsilon})$ for $\epsilon > 0$.

This requirement follows from Lemma 4.3. For the instance of 0-1-IP$^=$ to which we reduce has at most $f(r') \leq 2/3r$ variables and at most $g(k') \leq r$ clauses.

We store the resulting instance of 0-1-IP$^=$ in a $(2/3r + 1) \times r$ matrix M with polynomial-bounded entries, such that $M(i, j) = d$ iff d is the coefficient of variable i in constraint j, to which we add the result column.

From Remark 7.5 we obtain indeed that the bit representation of this kernel is indeed $r^{2-\epsilon}$ for some non-negative ϵ. $\square$

Corollary 9.2 (Bura [2]). coNP $\subseteq$ NP $\setminus$ P

Proof. Follows from Lemma 9.3, Theorem 9.1 and Lemma 9.2. $\square$

10. Conclusion

We have shown the mechanism through which a 1-3-SAT$^+$ instance can be transformed into an integer programming version 0-1-IP$^=$ instance with variables at most two-thirds of the number of variables in the 1-3-SAT$^+$ instance.

This was done by a straightforward preprocessing of the 1-3-SAT$^+$ instance using the method of Substitution.

We manage to count satisfying assignments to the 1-3-SAT$^+$ instance through a type of brute-force search on the 0-1-IP$^=$ instance.

The method we have presented before in the shape of Gaussian Elimination gives interesting upper bounds on 1-3-SAT$^+$, and shows how instances become harder to solve with variations on the clauses-to-variables ratio.

An essential observation here is that in this case this ratio cannot go below 1/3 up to uniqueness of clauses. This can be easily checked in polynomial time..

By reduction from 3-cnf-SAT any instance of 1-3-SAT in which the number of clauses does not exceed the number of variables is also NP-complete.

Our contribution is in pointing out how the method of Substitution together with a type of brute-force approach suffice to find, constructively, a non-trivial kernel for 1-3-SAT$^+$.

The most important question in Theoretical Computer Science remains open.

Acknowledgments

Foremost thanks are due to Igor Potapov for the support and benevolence shown towards this project.

Most of the ideas presented here have crystallized while the author was studying with Rod Downey at Victoria University of Wellington, in the New Zealand winter of 2010.

I am very much indebted to Noam Greenberg for supervising my Master of Science Dissertation in the year of 2012, one hundred years after the birth of Alan Turing.

I thank Asher Kach, Dan Turetzky and David Diamondstone for many useful thoughts on Computability, Complexity and Model Theory.

I have also found useful Dillon Mayhew's insights in Combinatorics, and Cristian Calude's research on Algorithmic Information Theory.

Exceptional logicians such as Rob Goldblatt, Max Cresswell and Ed Mares have also supervised various projects in which I was involved.

Western Australia is also in my thoughts, and I wish to thank Mark Reynolds and Tim French for teaching me to think, and act under pressure.

Special acknowledgments are given to my former colleague Reino Niskanen for many useful comments and for proof reading an initial compressed version of this manuscript.

References

[1] T. J. Schaefer, The complexity of satisfiability problems, in: Proceedings of the tenth annual ACM symposium on Theory of computing, ACM, 1978, pp. 216–226.

[2] V. Bura, The polynomial hierarchy collapses, arXiv 1808.02821.

[3] S. A. Cook, The complexity of theorem-proving procedures, in: Proceedings of the third annual ACM symposium on Theory of computing, ACM, 1971, pp. 151–158.

[4] R. M. Karp, Reducibility among combinatorial problems, in: Complexity of computer computations, Springer, 1972, pp. 85–103.

[5] L. A. Levin, Universal sequential search problems, Problemy Peredachi Informatsii 9 (3) (1973) 115–116.

[6] L. G. Valiant, The complexity of computing the permanent, Theoretical computer science 8 (2) (1979) 189–201.

[7] S. Toda, Pp is as hard as the polynomial-time hierarchy, SIAM Journal on Computing 20 (5) (1991) 865–877.

[8] M. Soos, Enhanced gaussian elimination in dpll-based sat solvers., in: POS@ SAT, 2010, pp. 2–14.

[9] M. Wahlström, Abusing the tutte matrix: An algebraic instance compression for the k-set-cycle problem, arXiv preprint arXiv:1301.1517.

[10] A. C. Giannopoulou, D. Lokshtanov, S. Saurabh, O. Suchy, Tree deletion set has a polynomial kernel but no opt^o(1) approximation, SIAM Journal on Discrete Mathematics 30 (3) (2016) 1371–1384.

[11] H. Dell, D. Van Melkebeek, Satisfiability allows no nontrivial sparsification unless the polynomial-time hierarchy collapses, Journal of the ACM (JACM) 61 (4) (2014) 23.

[12] B. M. Jansen, A. Pieterse, Optimal sparsification for some binary csps using low-degree polynomials, arXiv preprint arXiv:1606.03233.

[13] B. M. Jansen, A. Pieterse, Sparsification upper and lower bounds for graph problems and not-all-equal sat, Algorithmica 79 (1) (2017) 3–28.

[14] M. R. Garey, D. S. Johnson, Computers and intractability, W.H. Freeman, New York, 1979.

[15] R. Schroeppel, A. Shamir, A t=o($2^n/2$), s=o($2^n/4$) algorithm for certain np-complete problems, SIAM journal on Computing 10 (3) (1981) 456–464.

Graph Isomorphism is in P

Valentin B. Bura, M.Sc.

Abstract

We provide a polynomial-time algorithm for the Graph Isomorphism problem.

Searches are performed on two candidate graphs; they are based on a standard version of Breadth First Search.

The BFS searches return *traces*, which are collections of *skeletons*. Roughly speaking, a graph skeleton refers to the list of vertices in the order visited by the BFS.

To check if the candidate graphs are isomorphic it is sufficient to verify that the two returned traces contain identical skeletons, up to the order of skeletons.

Keywords. *Graph Theory, Computational Complexity, Graph Isomorphism.*

The Graph Isomorphism problem did attract a great deal of interest in the recent and not so recent past, commencing in the early [1, 2, 3], and late seventies [4, 5, 6].

In 1987 in [7] Uwe Schöning proves that if the Polynomial Hierarchy collapses then the class NP-Intermediate is empty and the Graph Isomorphism problem is NP-complete.

By late nineties we observe a turn in this area of research, due to the well understood by then practical significance of fast algorithms, perhaps in the vein of Schaefer in [5] who also considered special cases of highly general computational decision problems.

The shift in question proposed a focus on solving *special cases* of computational problems, and diverting attention from the general cases, since the more general question whether $P = NP$ was by then generally thought to be hard to answer rigorously [8, 9].

We note that, according to the New Scientist, an overwhelming majority of working mathematicians believed in 2012 that $P \neq NP$.

There does exist some weak evidence that this opinion survey was in fact performed in good faith at that particular point in time. Opinion, however, is highly far from constituting proof as one ancient saying goes, probably a highly obvious reference of this interesting piece of journalism [10].

More recently, in 2015, László Babai investigates the complexity of Graph Isomorphism [11], devising an algorithm with a significantly low upper bound. Professor Babai's result is highly special since it constitutes a serious attempt towards positive outcomes while focusing on the general case of a computational problem.

In [12] we show using some straightforward arguments from Linear Algebra that the Polynomial Hierarchy collapses to the level above $P = NP$.

We achieved this by studying 1-in-3-SAT and making use of the properties of this problem to encode its instances into systems of linear equations. These systems can be reduced in size efficiently using a type of redundancy check based on the method of Gaussian elimination.

Since the class NP-intermediate was thus shown to be vacuous, since the PH collapses, it follows easily from the statement of Schoning's result that Graph Isomorphism is NP-complete.

Hence, it is sufficient to provide a polynomial time algorithm for Graph Isomorphism in order to answer the $P = NP$ conjecture in the positive.

This is our main aim and preoccupation in the current article.

1. Preliminaries

Let $\mathbb{N}$ constitute the set of integers and let $\mathbb{N}^+$ be the set of positive integers. An unordered pair is any binary element $\{u, v\}$ with $u, v \in \mathbb{N}$. We write in shorthand $\{u, v\} \in \mathbb{N} + \mathbb{N}$.

As usual an (undirected) Graph G is a structure $G(V, E)$ where V is an unordered finite collection of objects labelled by $L \colon V \to \mathbb{N}^+$ such that $L[V] = \{1, 2, 3, \cdots, K\}$ with $K = |V|$, and $E \subseteq V + V$ is a finite collection of (unordered) pairs over V of the form

$$E = \{\{u, v\} \mid u, v \in V\}.$$

We say that two edges in E, $\{u_1, v_1\}$ and $\{u_2, v_2\}$ are *adjacent* if and only if either $L(u_1) = L(u_2)$, $L(u_1) = L(v_2)$, $L(v_1) = L(u_2)$, or $L(v_1) = L(v_2)$, that is, if and only if they hold one of their vertices in common. Two vertices u, v will thus be called adjacent if and only if there exists an edge $e = \{u, v\}$, that is, two vertices are called adjacent if and only if both vertices belong to some edge.

2. Breadth First Search

Let $G(V, E)$ be a graph with $L[V] = \{1, 2, 3, \cdots, K\}$ and $|V| = K$. Furthermore let $E = \{p_1, p_2, p_3, \cdots, p_r \mid p_i = \{u_i, v_i\}; L(u_i), L(v_i) \in L[V]\}$.

The algorithm BFS can be comprehensively described by "the sequential visit of each node of the graph G in a greedy fashion." More formally.

- Commence at (given) vertex $\bar{g}$ of G.

- Add $L(\bar{g})$ to ordered list T.

- Fetch the last vertex element g in list T.

- For each $e \in E(V)$ of the form $e = (g, h)$ add $L(h)$ to list T, iff $L(h)$ not already in T.

- Halt when $|T| = |V(G)|$.

3. Traces and Skeletons

A *skeleton* for a graph G is an ordered sequence s of vertices in $V(G)$ such that the following conditions are met:

- s is produced by applying BFS to G, and

- $|s| = |G|$, and

- s contains no repeated vertices.

The *trace* for a graph G is an ordered sequence S of skeletons such that the following conditions are met:

- there exist exactly $|S| = n$ skeletons in S where $n = |V(G)|$, and

- each member s of S is obtained by running BFS from a different vertex of G.

Note that a graph G with $|V(G)| = n$ will always exhibit n "different" skeletons. Note that any graph G will exhibit one and only one trace, up to re-arrangement and/or sorting of skeleton elements.

4. An algorithm for Graph Isomorphism

The algorithm is essentially described by the following sequential steps.

- obtain traces for graphs G and H.

- for each skeleton of G, check if there is a skeleton of H identical to it up to a relabel.

- if for each skeleton of G, there is exactly one skeleton of H identical to it (up to a relabel) return true, or else return false.

5. Algorithm correctness

Remark 1. *Let G, H be two isomorphic graphs. Then,*

$$\text{TRACES}(G) = \text{TRACES}(H)$$

Proof. Let $G \sim H$. Assume without loss of generality that

$$\text{TRACES}(G) = [[g_1^1, g_2^1, \ldots g_n^1], [g_1^2, g_2^2, \ldots g_n^2], \ldots [g_1^k, g_2^k, \ldots g_n^k]]$$

$$\text{TRACES}(H) = [[h_1^1, h_2^1, \ldots h_t^1], [h_1^2, h_2^2, \ldots h_t^2], \ldots [h_1^m, h_2^m, \ldots h_t^m]]$$

Also assume *without loss of generality* that $\text{TRACES}(G)$ and $\text{TRACES}(H)$ are presented as double-sorted with respect to indexing of vertices. This can be achieved at the expense of quasi-cubic ($n^2 \log n$) polynomial time.

Since $G \sim H$ we have that $m = k$ and $t = n$.

Also since $G \sim H$ and assuming $H = \alpha[G]$, commencing BFS at the node g in G with visited sequence $g_1, g_2, \ldots, g_s$ and commencing BFS at the node $\alpha(g)$ in H will result in sequence

$$\alpha(g_1), \alpha(g_2), \ldots, \alpha(g_s)$$

within the graph H. Therefore

$$\text{TRACES}(G) = \text{TRACES}(H)$$

up to isomorphism α. $\square$

Remark 2. *Let G, H be two graphs. Then, if* TRACES$(G) =$ TRACES(H), *then*
$$G \sim H$$

Proof. Suppose that TRACES$(G) =$ TRACES(H) up to isomorphism α. Then α itself witnesses the fact that $G \sim H$. □

Since the property $G \sim H$ can be checked in polynomial time we obtain the following result.

Theorem 3. *Graph Isomorphism is in P.*

Since we have in fact shown in [12] that the P.H. collapses, and provided our arguments are correct, then by the results of Uwe Schoning in [7] we have that Graph Isomorphism is NP-hard. On the other hand, by the arguments laid out above, we have shown that Graph Isomorphism can be checked in polynomial time.

Corollary 4. $P = NP$.

Acknowledgements

Noam Greenberg, Ed Mares, Rob Goldblatt, Max Cresswell, Dillon Mayhew, Carolyn Chun, Chris Atkin, Colin Bailey, Mark Reynolds, Tim French, Gernot Salzer, Miki Hermann, Reinhard Pichler.

Selected Bibliographical Material

[1] Stephen A. Cook, 1971.
The complexity of theorem-proving procedures.
citeseerx.

[2] Richard M. Karp, 1972.
Reducibility among combinatorial problems.
dblp.

[3] Leonid A. Levin, 1973.
Universal sequential search problems.
mathscinet.

[4] Larry J. Stockmeyer, 1976.
The polynomial-time hierarchy.
science direct.

[5] Thomas J. Schaefer, 1978.
The complexity of satisfiability problems.
dl.acm.

[6] Michael R. Garey,
David S. Johnson, 1979.
Computers and intractability.
dl.acm.

[7] Uwe Schöning, 1987.
Graph isomorphism is in the low hierarchy.
science direct.

[8] Mikhail Muzychuk, 2004.
A solution of the isomorphism problem for circulant graphs.
citeseerx.

[9] Samir Datta, Nutan Limaye, Prajakta Nimbhorkar, 2008.
3-connected Planar graph isomorphism is in log-space.
arXiv.

[10] New Scientist Magazine, 2012.
Poll consensus on million dollars logic problem.
web.

[11] László Babai, 2015.
Graph Isomorphism is in Quasipolynomial Time.
arXiv.

[12] Valentin B. Bura, 2018.
A kernel method for 1-in-3-SAT.
arXiv.